Mitsubishi Ki.67/Ki.109 HIRYU

In Japanese Army Air Force Service

by

Richard M. Bueschel

Schiffer Military/Aviation History
Atglen, PA

Cover artwork by Steve Ferguson, Colorado Springs, CO

TEETH OF THE DRAGON - Mitsubishi Ki-67 PEGGY - The Mitsubishi Type 4 heavy bomber was proof positive of the adage "too little, too late," owing that the *Hiryu* (Dragon Slayer) was the most adaptive attack bomber to serve both Imperial Army and Navy units. In the final desperate sea battles in their home waters, the PEGGY crews met their fate in the jaws of the American Navy dreadnaught.

Assigned to naval operations, the JAAF 98th Regiment was the first to employ their torpedo-armed Ki-67s against the American fleet raiding Formosa (Taiwan) in October or 1944. Iliustrated is a flight of 1st Chutai PEGGY's, the subject aircraft having just released its weapon within the withering anti-aircraft barrage of the enemy taskrorce.

Acknowledgments

Thanks are due to the photo sources cited, with special thanks to Richard L. Seely of Aero Literature for permission to reproduce the photographs from the R. L. Seely Collection; James F. Lansdale for proofing and editing of the original edition for historical accuracy, and James I. Long for enhancing and bringing the text up to date with current practice nomenclature, model and unit identification.

Book Design by Ian Robertson.

Copyright © 1997 by Richard M. Bueschel.
Library of Congress Catalog Number: 97-66914

All rights reserved. No part of this work may be reproduced or used in any forms or by any means – graphic, electronic or mechanical, including photocopying or information storage and retrieval systems – without written permission from the copyright holder.

Printed in the United States of America.
ISBN: 0-7643-0350-3

We are interested in hearing from authors with book ideas on related topics.

Published by Schiffer Publishing Ltd.
4880 Lower Valley Road
Atglen, PA 19310
Phone: (610) 593-1777
FAX: (610) 593-2002
E-mail: Schifferbk@aol.com
Please write for a free catalog.
This book may be purchased from the publisher.
Please include $3.95 postage.
Try your bookstore first.

Finest Japanese bomber of the Pacific War, the Mitsubishi Ki.67-Ia Type 4 Heavy Bomber, Model 1A, began to enter unit service in the summer of 1944. N. Saito.

MITSUBISHI Ki.67/Ki.109 HIRYU

To the pilots and crews of the newly-formed Japanese Army Air Force (JAAF) 110th Heavy Bomber Regiment at Hamamatsu on the east coast of the Japanese home island of Honshu, the raids seemed misplaced. They were completely over water. It was the kind of attack operation that the Imperial Japanese Navy Air Force (JNAF) had been trained to do since the middle 1930s. But now that the heart of Japan was coming under attack early in 1945 by the massive American B-29 Superfortress bombers from the Marianas—stationed almost 1300 miles away on Saipan, Tinian, and Guam—the JNAF was incapable of effectively striking back. The air forces of the Imperial Navy had all they could do to hold the line in the Philippines and elsewhere throughout Japan's scattered Pacific Island Empire. The defense of Japan had to be left largely to the Army. Any logical, or even illogical, attempt to hinder or halt the B-29 raids on Japan was a part of that defense. Thus, the Army found itself conducting bombing raids that took two days to accomplish, with most of that time spent in total darkness and completely out of sight of land.

The first sporadic B-29 bombing raids on Japan had originated from China in the summer of 1944. The American crews were inexperienced, and the ranges were too long to make any major impression. But the American invasion of the Japa-

The Ki. 67 was assigned the popular name Hiryu (Flying Dragon) by its manufacturer. In service it was more popularly known as the 4 Ju, or Type 4 Heavy Bomber. Sekai no Kokuki.

Normal Hiryu bomb load was eight 220 lb. Type 94 or three 551 lb. Type 92 bombs, for a total of around 1,700 pounds. Sekai no Kokuki.

nese occupied islands in the Marianas chain in July, 1944, changed all that. Japan's bastion at Saipan, long an aerial way-stop between Japan and the South Pacific, had fallen to the American enemy. Most rational Japanese regarded this as the beginning of the end, for now Tokyo itself was within bombing range. In July, three months before the first B-29 bomber touched down at the southern end of Saipan on 12 October, 1944, training was underway in the newly established 2nd Independent Flying Unit, organized at the JAAF's Heavy Bomber Instructing Flight Division at Hamamatsu in Japan to counteract the forthcoming American program. By the time the first Saipan-based B-29s appeared over Tokyo on 24 November, the specially modified bombers of the 2nd Independent Flying Unit had made two attacks at Isley Airfield on Saipan, and struck again the night after the ini-

Nose armament of the Ki.67-Ia was a flexible 12.7mm Ho.103 Type 1 MG. Sekai no Kokuki.

tial Saipan-based B-29 raid. In the meantime, an additional unit designated the 110th Heavy Bomber Regiment had also been formed at Hamamatsu with the same attack mission and had started to receive its original combat equipment in October. The aircraft type chosen to attack the American bases in the Pacific far to the south of Japan was the Mitsubishi Ki.67-I Type 4 Heavy Bomber, Model 1 Hiryu ("Flying Dragon"), the ultimate development of Mitsubishi's long line of Army heavy bombers. Regrettably serving in combat during WWII for only nine months, the Ki.67 bomber and its numerous variations, under the all encompassing Allied Code Name of "Peggy", became some of the best known Japanese aircraft of the closing months of The Pacific War.

The Saipan missions were grueling, but highly successful. The armed Hiryu bombers would leave Hamamatsu in the early evening for a three hour flight to Iwo Jima, a spot of volcanic land in the Bonin Islands almost 700 miles due south of Japan. The first destination was Chidori, the only air-

Hiryu dorsal turret position had a single 20mm Ho.5 cannon. Sekai no Kokuki. Sekai no Kokuki.

Power plants for all Hiryu production models were the powerful 18-cyl. 1,900 h.p. Mitsubishi Ha.104 Type 4 radial. Mannosuke Toda.

strip on Iwo Jima capable of handling bombers—it was officially known as Airfield No. 1—constructed on a flat plain of black volcanic ash facing the beach at the foot of Mt. Suribachi. There the crews would eat dinner and waste time while their aircraft were re-fueled and inspected for the longest leg of the mission. As dusk fell they would leave Iwo Jima, this time for a 600-mile flight to Saipan timed to arrive over the American airfields just as the giant B-29 bombers were being bombed and fueled for an early morning take-off against Japan. Coming in low over airfields, often only 400 feet off the ground, the Hiryu bombers would drop their loads and strafe the line of American B-29's, and disappear quickly into the darkness beyond, only to be faced with 2-1/2 more hours of night navigation and the need to get back to Iwo Jima before their short fuel supply ran out.

To the Americans, the raids were far more than a nuisance. Although they were briefly reported in the press at the time, the actual toll in lost American lives and aircraft was suppressed to avoid letting the Japanese know how successful their attacks had been. The peripheral toll in lost sleep, and reduced bombing efficiency by American crews over Japan in January and early February 1945, could not be counted. The Japanese seemed to have an uncanny ability to show up over Saipan at exactly the right time, hitting the American bombers at their most vulnerable moment before a raid. There was no question that the Japanese attacks had to be stopped.

From the Japanese point of view the raids were exorbitantly costly, but necessary. It was the only recourse they had. The overall operation was a very sophisticated one. High-flying Mitsubishi Ki.46-IIIa "Dinah" reconnaissance planes maintained a constant surveillance of the American bases. The massive preparations required for an upcoming B-29 raid were patently obvious from the air. The appearance of American weather-sniffing aircraft over Tokyo and Central Japan all but signaled a planned attack. The Hiryu bombing units reacted quickly, and would start on their island hopping journey that very afternoon. The Japanese losses over Saipan were substantial in themselves, for the JAAF bombers had to face a solid wall of anti-aircraft fire and aerial attack. The cost in men and materials was upped by the additional losses over the night-blackened Pacific as the hastily-trained green crews of the Hiryu bombers groped their way back to Iwo Jima and an airfield continually pock marked by American raids, making the price a dear one.

Soon after its first Saipan raid on the night of 6/7 December, when the 110th Regiment lost six of their eight aircraft involved, the 2nd Independent Flying Unit was merged with the regiment to add their combat experience and equipment to maintain unit strength. The raids were further increased when Ki.67 bombers of the 7th Heavy Bomber Regiment, then under JNAF control, also appeared over Saipan. As the Americans suffered, and the Japanese fed a continual flow of "Peggy" bombers into the program, both sides were strained to the limit.

The JAAF raids ended abruptly with the American invasion of Iwo Jima on 19 February, 1945. What had once been a Japanese advantage, in

Once necessary changes were worked out in the pre-production evaluation models, the Ki.67-I design was "frozen" to keep changes to a minimum in order to maintain high levels of manufacturing and delivery in 1944 and 1945. Mannosuke Toda.

Mitsubishi Ki.67/Ki.109 HIRYU

Hiryu crews were generally young and newly trained as a result of the earlier disastrous Army pilot losses in New Guinea. Sekai no Kokuki.

terms of an island refueling and emergency landing station between Japan and the Marianas, suddenly became a detriment and an American advantage over Tokyo. The Hiryu bomber that had just entered the war as the only answer to a current tactical problem was suddenly outranged. While the "Peggy" would remain the JAAF's most important bomber, ultimately operating over Okinawa and against Allied fleet units from bases in southern Japan and Formosa, its promise was never real-zed. As the war drew to an end, the most critical research and development task facing the 1st Army Air Arsenal at Tachikawa was increasing the range of the Ki.67 so the bombers could once again reach Saipan, otherwise American B-29 bombers would have free reign over Japan. Hastily modified examples of Hiryu bombers with additional fuel tanks were evaluated at the Flight Test Department at Fussa in the spring and summer of 1945 under the direction of Major Hideo Sakamoto, Bomber Test Commander, in addition to prototypes of an ultra-long-range Hiryu that had extended wings and enough fuel to range 5,000 km, (3,100 miles). But it was all too late. The aircraft unanimously regarded by Japanese and foreigners alike as the finest standard service bomber produced in wartime Japan was literally wasted in a fruitless effort to stem the losing tide.

The Creation of a Bomber

The magnitude and capability of Japan's wartime aircraft industry was nothing short of remarkable. In the brief decade prior to the Pacific War, Japan went from a modest aircraft producer relying on the borrowed technology of licensed-production to a position of proprietary excellence. In the heavy bomber field, most of this metamorphosis took place at Mitsubishi Heavy Industries, Ltd. In terms of people, Mitsubishi had more design engineers on the payroll experienced in the development of heavy bomb-

The business end of the 20mm cannon tipped dorsal turret of the Ki.67-Ia. Sekai no Kokuki.

Maintenance training for the Hiryu was at the Tokorozawa Instructing Maintenance Division, starting in June 1944. Hideya Ando.

ers than the rest of Japan's aircraft industry combined. This unique core group had been trained in Europe, the United States, and ultimately, in Japan, with two duplicated design sections completely separated to serve both the Imperial Japanese Army and Navy. By 1941 the leading service bombers of both the JAAF and the JNAF were Mitsubishi designs, with the exception of a short-term regression in the Army to a 1938 Ki.49 Nakajima design accepted for 1940 production as the Type 100 Heavy Bomber Donryu which finally entered service in 1941.

Late in 1940, prior to the recognition of general Army disappointment with the Nakajima Donryu, Army Air Headquarters began to think about its next generation replacement. At the beginning of 1941 the design specifications had been set down on paper and the Army was ready to tender development contracts. A rough draft outlined by Mitsubishi, with emphasis on production simplicity, led to Army Air Headquarters acceptance of the proposal and the 17 February 1941 assignment of the Ki.67 designation to the manufacturer for further work. It wasn't until the fall of that same year that Nakajima was invited to participate in the effort, with Army approval to proceed on a developmental project based on the Donryu under the Ki.82 designation. Where Mitsubishi had previously—albeit briefly—lost ground to Nakajima as the major JAAF bomber producer, the firm was now in the catbird seat. As Ki.67 development proceeded into the early 1940s, Mitsubishi's position became even stronger as Donryu production faltered and assembly of advanced versions of the Ki.21 Type 97 "Sally" bomber of the late 1930s was increased at the height of the Pacific War years. The Ki.67 became the hoped-for replacement for the Ki.21 while Nakajima's smaller and lighter Ki.82 re-design of the somewhat unremarkable Ki.49 came to its end in the planning

The Hiryu cowlings were highly streamlined, with fan-induced cooling Mannosuke Toda.

stages in the middle of 1942 at the firm's Mitak Research Division. By the end of 1942 Mitsubish had the field to themselves.

The design team for the Ki.67 project set up within the firm's Army Design Section combined the best of Mitsubishi's two technical worlds; the experienced and the imaginative. Responsibility for the overall project came under the jurisdiction of Fumihiko Kawano, chief technical officer of the Mitsubishi firm. Kawano had been the Chief Designer of the Ki.15, Ki.30 and Ki.51 reconnaissance and light bomber fixed gear monoplanes, and was the technical expert who had been most respon

Hiryu torpedo bombers of the 7th Air Regiment carried their unit number on the fin. Mounting shackles to carry the torpedo below the fuselage can be seen on the underside of aircraft 7-60 of the 3rd company. Mannosuke Toda.

The 7th Air Regiment torpedo bombers took part in the battle of the Philippine Sea, October 1944, as part of the "T-Force." This wrecked example of aircraft 7-77 from the 3rd Company was found in the Philippines. USAAF.

Modified at the Tachikawa Army Air Arsenal in August and September 1944, ten examples of Ki.67-I Special Attack bombers with plywood noses were completed in time for use by the Fugaku (Mt. Fuji) Special Attack Corps in the Philippines Campaign. This one was caught on the ground before it could be used. USAAF.

sible for moving the producer into the modern monoplane bomber field. Kawano assigned Hisanojo Ozawa, Mitsubishi's expert in Army bombers, the position as Ki.67 Chief Project Engineer. Ozawa had worked on all-metal monoplane bombers since the early 1930s, and was responsible for the design of the Mitsubishi Ki.2-II, the first JAAF bomber to have a retractable landing gear; served as co

The first use of the Fugaku Special Attack Corps Ki.67-I suicide bombers was on 7 November 1944, last on 12 January 194 Trigger-pole set off the bomb load on impact. Sekai no Kokuki.

Hiryu aircraft I-205 of the Fugaku Special Attack Corps in the Philippines warms up for its one and only mission, 7 November 1944.

designer of the highly successful Ki.21 Type 97 bomber series then in production; and led the redesign of the Ki.21-II model that made it possible to extend the service life of this venerable bomber well into the Pacific War period. Assisting Ozawa were two younger men who were working on their first Mitsubishi design; Teruo Tojo and Yoshio Tsubota, both trained outside of Japan and experienced in modern production methods. Tsubota was an Aeronautical Engineering graduate of the California Institute of Technology and had received most of his training in the United States just as the sec-

Fugaku Corps I-205 taking off. Bomb load was two 1,764 pound bombs, triggered by the nose pole. Sekai no Kokuki.

Mitsubishi Ki.67/Ki.109 HIRYU

Another Fugaku Corps Special Attack bomber takes off on its one way mission. Aircraft I-207 was flown by Captain Motoo Neki. Tail markings represented Mt. Fuji. Shorzoe Abe.

ond generation of American long range bombers, which ultimately saw service in the Pacific War, were being conceived.

The Mitsubishi contract called for the construction of three Ki.67 evaluation aircraft capable of carrying a single 500 kg. (1,102 lb.) bomb a radius of 700 km. (435 miles) at an altitude of up to 7,000 m. (22,965 ft.) with a top speed of 550 km./hr. (342 m.p.h). The design would also have to carry a bomb load of eight 100 kg. (220 lb.) or three 250 kg. (551 lb.) bombs. Crew was specified as 6 to 8, or even as high as 10 depending upon the design, with a minimum of three 7.92 mm and two 12.7 mm MG in nose and side gun positions, with the larger caliber guns in dorsal and tail positions, the latter a carry-over from the earlier Ki.49 specification. Power specifications were unusually unrestricted, with Army Air Headquarters willing to accept the use of two Mitsubishi Ha.101, Ha.104 or Nakajima Ha.103 radials. The 1,450 h.p. Ha.101 was already being specified for the advanced Ki.21-II Model version of the Type 97 Heavy Bomber, and Nakajima was still in the prototype stages of the 1,870 h.p. Ha.103, so Ozawa elected to proceed with the powerful 1,900 h.p. Ha.104, a fan-cooled 18-cylinder advanced development of the reliable 14-cylinder Ha.101. Experimental bench models of the Ha.104 had been running since 1940, and the engine promised to be dependable and ready for production by the time the Ki.67 design had gained acceptance.

The prototype airframe, serial 6701, was completed early in November 1942, almost a year after

Nose, tail and side gun positions of the Fugaku Corps Ki.67-I suicide bombers were faired over with plywood. Crew was reduced to three men. USAAF.

Mitsubishi Ki.67/Ki.109 HIRYU

When wrecked examples of the Ki.67-I suicide bombers were found in the Philippines, American Intelligence erroneously thought the faired over aircraft were special reconnaissance versions of "Peggy." The use of a full size aircraft as a flying bomb was inconceivable at the time. USN.

Available in greater numbers by the time of the Okinawa invasion in April 1945, the Hiryu became the prime Army bomber used in the campaign. Note typical JAAF "riding high" pilot sitting astride the cockpit while the aircraft taxis on the ground. Mal B. Passingham.

Wrecked Ki.67-Ia of 3rd Company, 60th Air Regiment, early 1945. USAAF.

60th Air Regiment marking is wide color band across fin and rudder. 3rd Company is yellow. USN.

the Pacific War had started, and after the problems of overseas bomber field maintenance had been forcefully brought to the attention of the JAAF. It was the best looking bomber ever produced in Japan, with long sleek lines unmarred by turrets or bulges, all gun positions being flush to the fuselage. Moved at night to the Mitsubishi test facilities at Kagamigahara Airfield, on 17 December, 1942 the prototype was taken up on its first flight by Mitsubishi staff pilot Oda. In February 1943 it was joined by the second experimental aircraft, serial 6702, and finally by the third, serial 6703, in April. little sensitive to the controls, and at 537 km,/h (334 m.p h) somewhat short of its hoped for top speed at its operating altitude the initial Ki.67 test models were regarded as perfectly acceptable beginnings toward new Army bomber, although requiring further development. By now the JAAF was sending bombers to Burma, the Dutch East Indies, and Rabaul, and felt the pressing need for newer equipment.

Eager to move development along, the JAAF gave Mitsubishi blanket approval for sixteen more evaluation aircraft, serials 6704 through 6719, incorporating a number of required as well as desirable

New Hiryu of the 60th Air Regiment caught on the ground, early 1945. Unit was one of the few to carry white combat stripe into 1945 service due to its assignment in China. Protective covers over engine nacelles. USN.

changes. Improvements included redesigned control systems and added fuel tankage to extend the range. Armament was beefed up by replacing the 7.92 mm Type 98 nose gun with a 12.7 mm Ho.103 Type 1 machine gun in an improved nose position that was shorter and structurally stronger. Similar structural redesign work was carried out on the tail and side gun positions. The original single 12.7 mm machine gun was retained in the tail and the 7.92 mm guns remained in the port and starboard beam positions. The first of these improved Ki.67 evaluation aircraft was completed in May 1943; the last in March 1944, nine months later. By this time the JAAF had approved the Ki.67 for production as a standard service bomber. The production model was to be called the Type 4 Heavy Bomber Model 1A Hiryu, known as the Ki.67-Ia. It soon became known in Japanese military jargon and official abbreviations as the 4-Shiki Ju, or Type 4 Heavy. It was the last heavy bomber to enter series production for the JAAF.

More often than not caught on the ground by raiding Allied aircraft, the Hiryu bombing force never really had the opportunity to prove its merit. USN.

A moment of rare repose. 74th Air Regiment on Formosa in May 1945, just after receiving its new Hiryu mounts. Mannosuke Toda.

An Excess of Evaluation

In spite of its late arrival on the wartime scene, the Ki.67 Army Experimental Heavy Bomber, known even in its prototype stages as the Hiryu, was produced, modified and tested in a greater variety of models than any other JAAF heavy bomber. Most of this work was done with the sixteen improved Ki.67 evaluation aircraft. The primary thrust was for an acceptable Ki.21 Type 97 bomber replacement as the Ki.67-Ia, the design of which was frozen on 2 December, 1943, in order to proceed with production at three Mitsubishi plants, as well as at Kawasaki, Nippon Kokusai and at the 1st Army Air Arsenal at Tachikawa. These three additional manufacturers had large-aircraft productive capacity that could be immediately utilized to rush the Hiryu into service to meet the burgeoning needs in the South Pacific, New Guinea, Burma and China. By the time the first production models began to come off of the lines at Mitsubishi's Plant No. 5 at Nagoya in April 1944, the war fronts had changed drastically and Japan was only a few months away from aerial attack by American B-29 Superfortress bombers.

Production Ki.67-Ia aircraft, beginning with airframe 20, were similar to the early Ki.67 evaluation aircraft, with armament further increased by changing the side blister hand-held 7.92 mm Type 98 MGs to 12.7 mm Type 1 Ho.103 arms, an armament format that remained standard on the frozen production model until the completion of airframe 451 in May 1945, at which time the tail turret position was increased to two 12.7 mm Type 1 Ho.103 MGs, with

Unit marking of the 74th Air Regiment was white, with a unit device made up of a contraction of the numbers "7" and "4." This is aircraft number 148 of the 1st Company. Mannosuke Toda.

the aircraft continuing in production as the Ki.67-Ib, Model 1B.

Under normal conditions the production of the Hiryu should have been handled with ease. The aircraft had been designed for mass production; virtually all of the Hiryu lines were straight and did not require complicated jig work; manufacturing facilities were standing; and the older Ki.21 Type 97 was taken out of production at Mitsubishi's No. 5 plant to make room for the newer bomber. But the Hiryu production program was plagued with problems. The American bombing effort, modest at first, began to eat away at production efficiency, particularly in terms of employee absenteeism. In December 1944, the devastating earthquake that struck Japan severely hampered Ha.104 engine production, forcing an immediate drop in Hiryu completions. And then problems that need not be came up, with poor production management and inefficient machine tool utilization, rising personnel problems and conflicts, and a thoroughly confused plant dispersal program. This was all backed up by the addition of an utter confusion of change orders as the JAAF kept revising the bomber to adapt to the ever-changing war conditions, an impossible task in itself. The result was a start-and-stop-and-start production effort that kept Hiryu production down to a total of about 688 aircraft of all models by the time the war ended in the middle of August, 1945. During this period the JAAF had planned for the production of 1840 of the bombers at Mitsubishi alone.

The sub-contracting effort was even more disappointing. Kawasaki was scheduled to produce 592 of the bombers by the end of 1945, starting at the firm's Kagamigahara works in December 1944. Only 81 were completed, with Mitsubishi supplying fuselage assemblies while Kawasaki made the wings and tails and put the aircraft together. The effort at Nippon Kokusai was even worse. In December 1944 Mitsubishi sent 29 fuselage assemblies and some wings to the Kokusai plant. The aircraft were barely put together before the end of the war. At the 1st Army Air Arsenal at Tachikawa a single Hiryu was assembled from Mitsubishi components, but no more.

The Ki.67-Ia Hiryu that entered JAAF service in the summer of 1944 gave the Imperial Japanese Army a better bomber than the Navy for the first time since the JAAF Ki.1 and Ki.2 bombers of the early 1930s. The Hiryu almost handled like a fighter, and was the first modern Japanese bomber that could be looped or thrown into a vertical turn without a terrible struggle to maintain control. Self-sealing fuel tanks, crew armor, and a porcupine-gun

74th Air Regiment re-equipment, Ki.67 Hiryu. Summer 1945. Mannosuke Toda.

Ki.67-Ib Hiryu of the 74th Air Regiment on Formosa. Armament has been revised to carry two 12.7mm Ho.103 Type 1 MG in the dorsal turret. Mannosuke Toda.

Cockpit of the Mitsubishi Ki.67-Ib Hiryu Type 4 Heavy Bomber, Model 1B. Mannosuke Toda.

defense gave the crew a reasonable chance for survival. Had the JAAF been able to make use of the experienced crews that had been literally thrown away at Rabaul and New Guinea, the Hiryu would have been a far more formidable opponent than it turned out to be. But the hot new bomber arrived at a time when the JAAF had suffered grievous losses, and crew training at the Hamamatsu Heavy Bomber Instructing Flight Division had to literally start from scratch in the summer months of 1944.

The preparations to get the Hiryu ready for combat were just as diverse as the variety of models being produced. The initial confusion of Hiryu "modelitis" was felt by the Army Air Test Department. Under the command of Lieut. General Ogata, the Air Examination Division at the JAAF's massive experimental and evaluation base at Fussa found itself with the Hiryu becoming its number one priority. Of the nine pilots assigned to bomber testing, all but a few worked on the various Hiryu developments, with Major Hideo Sakamoto placed in command of the program. Time was of the essence. Before the Pacific War the JAAF had allocated up to two years to test a new bomber model. With the arrival of the Ki.67 evaluation aircraft the Experi-

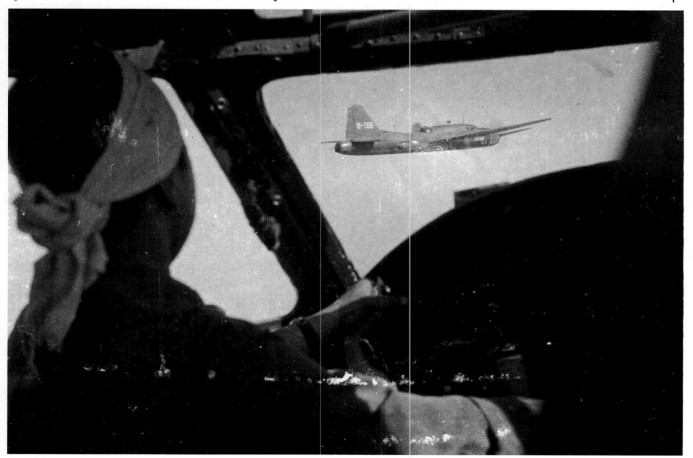

74th Air Regiment in the air over Formosa with its Hiryu bombers. Aircraft 166 in the distance. Mannosuke Toda.

Hiryu bombers of 1st Company, 74th Air Regiment. Aircraft 146 next in line. Mannosuke Toda.

mental Section of the Air Examination Division was instructed to complete the task in six months. It became a night-and-day task, with Major Sakamoto coming to regard the Ki.67 Hiryu and the parallel Tachikawa K.74 Army Experimental Long Range Bomber projects as his own "beloved children."

One of the first priorities at Fussa was to convert the internally-loaded Ki.67-Ia heavy bomber to a torpedo bomber as an anti-invasion weapon. The idea had first been suggested early in the Hiryu's life in December, 1942. On 5 January, 1943, the JAAF Purchase Orders tendered to Mitsubishi called for the addition of externally-mounted torpedo racks on a run of one hundred of the bombers. In order to evaluate the new configuration, two early Ki.67 evaluation aircraft, serials 6717 and 6718, were modified to mount the racks, and shipped off to Fussa for tests. Major Sakamoto, faced with the use of Navy weapons on an Army aircraft, promptly took the two bombers to the Naval Air Test Center at Yokosuka early in 1944 in order to conduct drop tests over Tokyo Bay. The evaluation procedure was complicated, for JAAF

Hiryus of the 3rd Company, 74th Air Regiment, Formosa, summer 1945. Mannosuke Toda.

22 Mitsubishi Ki.67/Ki.109 HIRYU

After flying the Nakajima Ki.49 Donryu in the Philippines, reassignment to Formosa and Japan and the arrival of the new Mitsubishi Ki.67-Ib Hiryu was a quantum upgrade for the 74th Air Regiment. Mannosuke Toda.

The 74th Air Regiment survived the war as a unit, to be disbanded at Matsumoto, Japan, after the war had ended. Sekai no Kokuki.

pilots were being asked to master a naval attack procedure that even JNAF pilots found difficult. The tests continued into the summer, and by August Army Air Headquarters had been sufficiently impressed with the results to order Mitsubishi to fit torpedo racks to all Hiryu aircraft commencing with airframe No. 161. The JAAF command also made provisions to establish two Army torpedo bomber regiments and release a substantial number of Hiryu aircraft to the JNAF for use as a naval bomber.

The availability of a new airframe on the verge of mass production led to an analysis of other possible uses for the Hiryu. An obvious choice was as a high-speed glider tug for a massive new troop glider then under development for Imperial Army assault troops. The Kokusai Ku.7 Army Experimental Transport Glider was larger than any previous glider developed for Army service, and needed tow

Mitsubishi Ki.67/Ki.109 HIRYU 23

The Hiryu was produced by more manufacturers than any other JAAF bomber, with assembly by Mitsubishi, Kawasaki, Nippon Kokusai and the Rikugun Army Air Arsenal at Tachikawa. Mannosuke Toda.

aircraft more powerful and faster than the available modified Ki.21-IIc Type 97 tugs originally considered. The giant Ku.7 was first conceived late in 1942 as an invasion transport to carry 32 fully-armed troops, or an 8-ton tank, to the battle site. By the time the all-wood glider—the largest to be built in wartime Japan—was first completed in prototype form in March 1944 the war situation had altered the specified mission. The Ku.7 became an assault glider to carry troops and equipment to weak points in the Japanese defense perimeter and halt the Allied advance. It was accepted for limited production in the summer of 1944 as the Ku.7-I Transport Glider, Model 1 Manazuru even prior to its first test flight. The decision was in part the result of work already underway at the 1st Army Air Arsenal to convert a number of Ki.49-IIa Donryu bombers and the evaluation Ki.67 serial 6713 to glider tugs, with the Hiryu modifications completed in February. On 15 August, 1944, the glider and its Hiryu tug made their joint first flight. The test was adjudged successful, and development proceeded as the war moved on, although the duo was never destined to see combat. The Ku.7 glider was code named "Crane," but was never met in action.

A late-comer to the Hiryu development program was a far more dramatic application, for it guaranteed the destruction of the aircraft and its crew on every mission. Long before the Kamikaze ("Divine

Modified as "Mother" planes for the I-Go series of anti-shipping missiles, ten Hiryus served at the Army Air Test Department at Fussa in this capacity beginning in November 1944 until the end of the war. Sekai no Kokuki.

Mitsubishi I-1a missile under test at Fussa attached to a modified Ki.67-I Hiryu. Type 4 Bomber "Mother" aircraft also tested the Kawasaki Ki.148-I-1b and the Tokyo Imperial University I-1c early in 1945. Mannosuke Toda.

A standard Ki.67-Ia Hiryu with towing equipment lifts off with its tow, the Kokusai Ku.7 Army Experimental Transport Glider. Hideya Ando.

Wind") concept of suicidal death in battle in order to destroy the enemy became a part of the Allied vocabulary, arguments were taking place at Army Air Headquarters in Tokyo over the advisability of pursuing such a program as a last-ditch measure to strike at Allied invasion vessels while still at sea in order to end the string of Allied victories. The unequivocal and final loss of New Guinea in June 1944 strengthened the hand of radical JAAF officers, proposing the use of Army bombers as flying bombs, guided to their targets by volunteers willing to die for the salvation of the Japanese Empire. After the plan was persuasively put forward by Army Air Headquarters staff officer Shigeru Ura, authorization to proceed with the development of suicide bombers to equip the first four JAAF units of the newly formed Special Attack Corps was given soon thereafter. By August examples of both the Nakajima Ki.49-IIa Donryu and Mitsubishi Ki.67-Ia Hiryu were being secretly modified by their respective manufacturers to carry internal bomb loads triggered on target contact by a fuse pole extending from the nose of the aircraft. The Hiryu modification, destined to be flown by the Fugaku ("Mt. Fuji") Special Attack Corps, was envisioned as a weapon twice as powerful as the modified Donryu, for it was designed to carry two 800 kg. (1,764 lb.) bombs a great distance out to sea. The Special Attack Corps Hiryu bombers were modified in September, with all turrets removed to be faired over with plywood to increase streamlining and speed. The crew was reduced to 3, the minimum necessary to navigate, maintain radio contact, and fly the aircraft. Activation of the bombs was an automatic reaction to impact. Following testing by the Air Examination Division at Fussa, by the end of September ten of the Ki.67-I Special Attack Bombers were available

First flown 15 August 1944, the Kokusai Ku.7 could carry an 8-ton tank or up to 32 fully armed troops. Hideya Ando.

Trailing behind its Hiryu tow aircraft, the Ku.7 Manazuru (Crane) was the largest glider built in wartime Japan. Mannosuke Toda.

26 Mitsubishi Ki.67/Ki.109 HIRYU

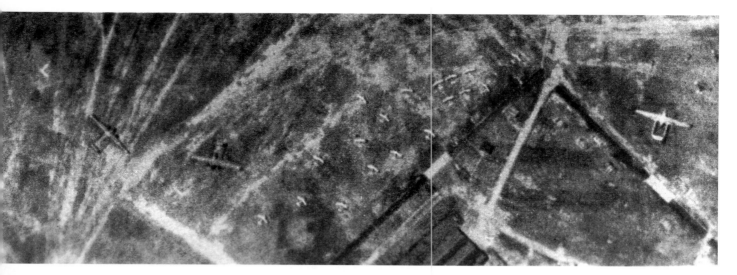

Spotted on the ground by Allied reconnaissance aircraft in spring 1945, the Ku.7 was thought to be a Japanese version of the German Gotha Go.242. Ku.7 glider at left, with two powered Ki.105 transport versions at right. USAAF.

for operations with two more soon in hand. The Fugaku unit formed by the 1st Flying Training Unit of the Hamamatsu Training Air Division was trained and ready to go into action by the end of October with a strength of 26 men and twelve specially modified Hiryu suicide bombers.

Evaluation by Operation

The long awaited combat arrival of the Hiryu bomber was now at hand, and the timing could not have been any tighter. Just as the Hiryu torpedo bomber units were being trained, and the Fugaku Special Attack Corps received its modified aircraft in the middle of October, the invasion of the Philippines began. Only one month earlier the JAAF had made elaborate plans to reinforce the Philippines in anticipation of an invasion, with re-assignment of heavy bomber regiments from elsewhere in the empire to the Clark Field area. At that time no conventional Army Heavy Bomber Air Combat Regiments were yet equipped with the Ki.67-Ia Hiryu bomber. But the special torpedo bomber units were under training, and once Allied fleet units began moving into the area, these were the first Hiryu units to go into action.

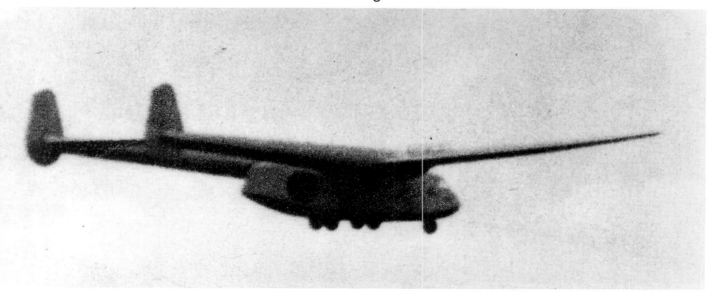

The Kokusai Ku.7 could be towed by the Nakajima Ki.49 Donryu or the Mitsubishi Ki.67-Ia Hiryu, although the Mitsubishi bomber was preferred due to its greater power and faster tow speed. Hideya Ando.

Mitsubishi Ki.67/Ki.109 HIRYU

Hiryu towing tests of the Ku.7 indicated stabilizer changes to improve tow and glide performance. Hideya Ando.

The official Imperial General Headquarters Sho 1 plans for the defense of the Philippines dated 22 September, 1944, stated that raiding enemy task forces were to be attacked only by designated air units, all of which were in the JNAF, "with the exception of some Army Air Units equipped with Type IV Bombers and undergoing special training in Japan proper." The units mentioned made up the mobile Taifu ("Typhoon") Attack Force, named for its all-weather attack capability. Shortened to "T Force", these units were being specially trained to attack American aircraft carriers at night or under difficult weather conditions with their modified Hiryu Type 4 torpedo bombers. Under a single command the "T-Force" could be moved anywhere as a special force to attack Allied invasion attempts. Primarily naval aircraft, the force included the 7th and 98th Air Regiments, both of which were old-line JAAF Heavy Bomber regiments that had been placed under Imperial Japanese Navy command when they were attached to the JNAF 2nd Air Fleet as part of the Philippines defense force on 25 July 1944. In addition, torpedo bomber versions of the Hiryu had been delivered to the JNAF in August to equip the 762nd Naval Air Corps, also attached to

Prototype Kokusai Ku.7 comes down after test flight. The large glider was code named "Buzzard." Hideya Ando.

Late model Ki.67-Ib bombers were radar-equipped to seek out Allied shipping in dirty weather. Hideya Ando.

the 2nd Air Fleet and composed of the Navy's crack bomber pilots. Known as the Yasukuni Naval Air Group, the JNAF Hiryu unit was named after Japan's most honored Shinto Military shrine. All of the Hiryu units were scheduled to be moved to southern Formosa in September to put the "T-Force" within striking distance of American carriers approaching Formosa or the Philippines. Due to training and equipment delays only a portion of the striking force had reached Formosa by early October. The main strength of the JNAF "T-Force" remained in Japan on southern Kyushu. Supporting this force were fifty JAAF Hiryu torpedo bombers of the 7th and 98th Air Regiments.

As Task Force units of the USN 3rd Fleet approached Luzon and southern Formosa in the second week of October, 1944, briefly striking at Okinawa, an attack warning was issued by Combined Fleet Headquarters in Tokyo. On 10 October the land-based air forces of the JNAF 2nd Air Fleet—including the two Army Hiryu regiments—were put on combat alert. On 12 October American carrier bombers and fighters struck at Formosa and its surrounding islands. The time had come, and the aerial phase of the Battle of the Philippine Sea was underway.

Japanese reconnaissance reported that three American carrier groups were operating off the east coast of Formosa. As darkness fell on the evening of the 12th, JNAF "T-Force" bombers left Kanoya Air Base on Kyushu, attacked an American Task Force, and returned to airfields on Formosa to report the sinking of four carriers, with ten other ships set afire. The Army Hiryu torpedo bombers, flying

Radar "whiskers" were also on the fuselage, just forward of the side gun positions. Sekai no Kokuki.

Mitsubishi Ki.67/Ki.109 HIRYU

Hiryu wreckage littered Formosa and Japan when The Pacific War was over. Seiso Tachibana.

from Kyushu and Okinawa, also made a contact and reported the sinking of two unidentified ships, possibly carriers. The next night the attacks were repeated by 32 "T-Force" aircraft with two more American carriers reportedly sunk, and a third left in flames. The next day there was a brief morning attack, followed by an afternoon attack by 124 "T-Force" bombers against another American Task Force, this time leading to the claim that another carrier was hit, in addition to three cruisers. The Army Hiryu torpedo bombers came back into the battle as night fell, with a force of 70 mixed bombers claiming the sinking of another two American carriers. It appeared that Imperial Japan had won the greatest naval victory of the war. There were celebrations in the streets of Tokyo, Yokohama, Kobe and other major cities in Japan as the news was announced. The populace cut loose after months of depressing news. There was even talk of ending the war in victory.

The reported battle successes were false. They never really happened. Japan's euphoria quickly turned to a deep depression as reality replaced fiction. The facts were far worse than imagined and were never reported to the Japanese public. Of the twenty Hiryu level and torpedo bombers launched by the 98th Regiment on 12 October, eleven didn't return. Of the sixteen sent out two nights later, only one returned, three crash landed and twelve were lost in the action. The 98th Regiment lost twenty-six of its force of twenty-eight aircraft in two days destroying the viability of the unit. When USN aircraft from the American Task Forces that were reportedly destroyed by the "T-Force" struck at Luzon on the 15th, followed by the sighting of thirteen carriers in three Task Forces by JNAF reconnaissance aircraft on the 16th, the errors in the enthusiastic "T-Force" combat reports became obvious. The JNAF 2nd Air Fleet had thrown its full force at the American 3rd Fleet, and had hardly scratched it. In the process, most of the Hiryu bombers committed to the battle had been lost. In a matter of days the Allied invasion of the Philippines was un

Type 4 Heavy Bomber crew entrance was by nose hatch, revealed in this post-war wreck. Sekai no Kokuki.

...lose hatch door from the side. Sekai no Kokuki.

When hostilities ended, the surviving Hiryu bomber fleet was disarmed by the removal of propellers and guns. Mannosuke Toda.

...derway, and Japanese air power could do nothing to stop it.

With the Allied landings at Mindoro in November the Hiryu once again led the way into the next phase of fighting. The failure of the "T-Force" in October had demonstrated that conventional tactics against Allied surface vessels, the latter protected by swarms of fighters, were all but useless. At the end of October the Fugaku Special Attack Corps reached Luzon, barely too late to take part in the air offensive of the 24th and 25th of October. Rather than waste the suicide bombers on minor missions, the commanding 4th Air Army decided to hold the one-way bombers in reserve for an important strike. It was wise to wait, but also costly, for the modified bombers came under heavy aerial attack. American pilots were spotting "Peggy" for the first time in combat, and the bomber made an important target. The Fugaku unit was scheduled for twelve Special Attack Hiryu bombers at full strength, but only ten were produced in September and turned over to the unit. Ground losses in the Philippines cut this further, and when the unit was finally committed to battle against the supply shipping supporting the Mindoro invasion on 7 and 13 November, only about half a dozen of the aircraft were available for use. Their last operational action was on 12 January 1945.

The results were so successful in terms of aircraft expended for ships sunk or damaged, to the Japanese the Kamikaze attack became a completely acceptable form of warfare, coming to full flower in the Okinawa invasion five months later. To the American ground forces reaching the Manila area, the wrecked solid-nose "Peggy" bombers without gun turrets left behind by the Fugaku unit posed a mystery. It was speculated that these were fast, unarmed reconnaissance versions of the

Intact examples of the Mitsubishi Ki.67-I were acquired for evaluation by Allied intelligence units. Imperial War Museum.

Hiryu, although the make-shift plywood panel patching was confusing. The idea of using large twin-engined bombers by the Japanese to make a death dive into enemy naval vessels had not yet sunk into the American sensibilities.

Bomber Models and Mods

The crushing defeat of Japanese air power in the Philippines gave the JAAF a lot of problems, but it also gave it a rest. The only areas where the opposing air forces were in constant and direct confrontation were in Burma, China, and over Japan. Hiryu production, barring delays due to bombings and the great earthquake of December 1944, continued, and the regimental re-equipment program proceeded. The 60th Heavy Bomber Regiment, one of the oldest bomber units in the JAAF, took the Hiryu to China near the end of the year. Of greater

Hasty overpaint of U. S. insignia over the red Japanese hinomaru markings. Sekai no Kokuki.

Captured Japanese aircraft, including the Hiryu, were assembled at Yokosuka in October 1945 for trans-shipment to the United states for test and evaluation. Many were lost at sea when the two escort carriers bringing them to America were caught in a typhoon. Strapped to the decks, the aircraft were dumped overboard. Some day they will be found by divers. Robert C. Mikesh.

significance, the 2nd Independent Flying Unit and 110th Regiment were formed at Hamamatsu, receiving their Hiryus in July and October 1944, to bomb Saipan and other American occupied island B-29 bases in the Pacific. By the time Okinawa was invaded in April 1945, the Hiryu had further equipped the 16th Air Regiment in Chosen (Korea) and Okinawa; the 62nd Air Regiment in Japan to add to the Saipan bombing effort; the 95th Air Regiment in southern Kyushu; the newly formed 107th Heavy Fighter Regiment, a unit flying a few examples of the Hiryu to be prepared for later delivery of interceptor models of the twin-engined bomber; followed by the 14th, 61st, and 74th Heavy Bomber Regiments. Unit conversion training was handled by the Hamamatsu Heavy Bomber Instructing Flight Division in Japan and the 28th Flight Drilling Company at Ipoh in Malaysia, the latter flying both the Ki.54b trainer and Ki.67-Ia bomber.

The Okinawa campaign saw the introduction of the first major Hiryu model change, with the Ki.67-Ib entering service in March 1945. The visible difference was an increase in defensive armament with the addition of another 12.7 mm MG, although the internal differences varied periodically as new equipment was added, tested, dropped, improved, and added again. When the anti-invasion Ten-Go operation was staged out of Kyushu to attack Allied fleet units approaching Okinawa on 6 and 7 April, substantial numbers of the newer "Peggy" bombers were in the force. They remained the most important JAAF bomber type in service during the attacks of April and May. Flying out of Kyushu, later models mounted radar to detect Allied ships, with radar installations and even a searchlight installed as anti-fighter defense equipment. Later equipment also included an electric altimeter for low level attacks, with the Hiryu ending its combat days coming in low, attacking in one fast pass, and getting out of the target zone before the ever-present gaggle of USAAF, USN, RAAF, RAF and other Allied fighters could crowd in for the kill. The only thing missing in the Hiryu's combat capability was an adequate range and bomb load. This was on the way to being solved, for production of the Ki 67-Ic Model 1C with a 66% increase in bomb load to

Mitsubishi Ki.67/Ki.109 HIRYU

Last combat version of the Hiryu series, the Ki.109 Army Experimental Interceptor mounted a 75mm Type 88 cannon in its nose as a B-29 interceptor. Hideya Ando.

1,250 kg. (2,756 lb.) was already scheduled for the summer of 1945, to start with airframe No. 751. Production slow-downs delayed the change, and the production Model 1C "Peggy" was never built. The extension of range was an equally difficult task, for the aircraft already had a high level of design economy, leaving little room or weight-carrying capacity for additional fuel. While one group at the 1st Army Air Technical Arsenal was increasing the bomb load to create the Ki.67-Ic, another group was working in the opposite direction to reduce the bomb load and increase the range to maintain the bombing program against Guam, Tinian and Saipan in the Marianas. Twelve examples of a long-span Hiryu were being tested in the summer of 1945, but the critical delivery delays with the standard Ki.67-Ib Hiryu kept the newer model out of production. A parallel project was the development of alternate fuels to keep the JAAF flying in the face of a crunching fuel shortage. Power plant engineers at the 2nd Army Air Technical Research Institute had developed an alcohol fuel out of potatoes and sugar that worked fairly well on tests, and powered most JAAF training aircraft by the end of the war. Hiryu Ki.67-Ia airframe No. 52 had been converted to the highly corrosive alcohol fuel, with aluminum fuel tanks replacing the rubber tanks normally used for gasoline. The modified aircraft of unknown designation was awaiting flight tests at Kofu when the war ended.

While the conventional uses of the Hiryu commanded most of the Ki.67 production and developmental time, the aircraft was selected for a grow-

The only combat unit to receive the Ki.109 was the 107th Air Regiment at Micahara, at Kumagaya, in Japan. 14th Air Regiment was backup unit. Hideya Ando.

Production of the Ki.109 was 22 aircraft, experimental and operational. They were never able to intercept the B-29 due to poor climb and low ceilings. U. S. markings applied for test, but never pursued. Aircraft destroyed in Japan. USAF.

ing list of unconventional applications. The success of the Hiryu as a Special Attack bomber had already been proven by the Fugaku Corps in the Philippines in November, and an additional five examples of the aircraft were produced by Mitsubishi in December 1944. It was about this time that the JAAF received technical data on the German "Cherry Bomb", a lens-shaped bomb designed to focus its maximum explosive charge forward at the focal point. The diameter of the German bomb was virtually identical to that of the Hiryu fuselage interior, permitting the mounting of the charge behind the pilot's seat much like a back wall made of explosive, bursting forward through the pilot on impact. In its final form the new Special Attack Bomber had a crew of two and carried a single 2900 kg. (6,393 lb.) Sakura-Dan Bomb. Two of the aircraft were modified from standard Hiryu bombers by the Mitsubishi experimental shops at Nagoya in February 1945, with the work closely supervised by the Air Examination Division at Fussa.

Due to its availability and speed, the Hiryu was also selected as the "Mother Plane" for two of the guided glide bombs under development in the closing months of the war designated in the Ki.140-series. The first winged bomb in the series was the Mitsubishi Type I Glide Bomb, Model 1A, developed as an anti-shipping weapon designed to be carried to within seven miles of its target by a modified Ki.67-Ia bomber fitted with radio flight control instrumentation. Known in abbreviated form as the I-Go-IA, work on the glider bomb was initiated in the

Research Department of Mitsubishi Experimental Works No. 1 under the supervision of Research Director Sueo Honjo after receiving a JAAF development contract in July 1944. Carrying a 800 kg. (1,764 lb.) bomb on stubby wings, the glider was powered by a chemical rocket engine that provided 75 seconds of thrust. Stabilized by a gyro device connected to the horizontal tail surfaces, the bomb was visually guided to its target by a controller in the Hiryu "Mother Plane" who had control via radio over the elevator and rudder of the bomb. The first I-Go-IA bomb was completed in October 1944, test flown in November, and the device was being readied as a combat weapon in the summer of 1945.

Another free flying glide bomb with a stubby wing on a long fuselage was under development at Tokyo Imperial University in the spring of 1945 under JAAF supervision as the Rikagun Type I Glide Bomb, Model 1C, or I-Go-IC. Its Chief Designer was Professor Hideo Itokawa, former Design Chief of Nakajima fighter aircraft, who was now head of the Aeronautical Engineering Design Department at the University. The I-Go-IC was designed to home on the center of a shock wave pattern, specifically the naval guns fired by an invading fleet. As the guns were fired, the I-Go-IC would zero-in on its target, and impact at the gun turret. Three fairly successful flight tests were made with a Hiryu "Mother Plane" in March 1945, and twenty of the flying bombs were hand assembled by July to pave the way for mass production. As many as ten Hiryu bombers were reportedly modified as "Mother Planes" for the I-Go-IA and I Go- IC test and evaluation programs, and to be ready for eventual combat use.

Work had also progressed to the preliminary prototype stage on an advanced standard bomber model of the Hiryu to be produced in 1946 as the

Second experimental aircraft serial 10902 was modified at Fussa early in 1945 to increase altitude performance with the Ru.3 turbo-supercharged Mitsubishi Ha.104ru radials of 1,900 h.p. to serve as the pattern for the Ki.109-II series. Hideya Ando.

The surviving Ki.109-I aircraft were found by the occupying Allied forces at Micahara prior to their conversion to Ki.109-II standards. Holmes G. Anderson.

Ki.67-II Model 2. The project was initiated on 2 December, 1942, when the decision was reached to power an advanced model of the Hiryu with the powerful 2,400 h.p. Mitsubishi Ha. 214 radial, then in the design stage. A few of the engines were assembled in 1943, but problems so plagued the project that only sixteen more had been completed by the summer of 1945. As design work progressed on a strengthened Ki.67-II airframe with additional fuel tanks for extended range, Ki.67 pre-production evaluation aircraft serials 6716 and 6717 were fitted with experimental models of the Ha.214 power plants for flight testing. They were fitted with 6-bladed propellers. Majors Sakamoto, Oura and Otsuka flew the test bed bombers in tests at Fussa, hoping to successfully complete the evaluation before the end of 1944. The early promise shown by the re-engined "Peggy" bombers led to Ki.67-II prototype approval. By the time the Pacific War came to an end two Ki.67-II examples were under construction at Mitsubishi's No.1 Experimental Airframe Works. The first example was scheduled for completion in June 1945, although Allied bombings greatly delayed progress. The aircraft was only half finished when the Japanese surrender was announced. The switch to production of the Ki.67-II would have taken place in the summer of 1946, with the new Hiryu Model 2 becoming the JAAF's last standard bomber and a more powerful tug for the large assault gliders then under development.

Bomber to Fighter and Beyond

The idea of creating a defensive fighter out of the Hiryu airframe occurred early in the aircraft's career. At the time the Ki.67 Heavy Bomber designation was first assigned in February 1941, a parallel bomber escort version for Hiryu formations was specified as the Ki.69 project. Actual experience with similar conversions of the Ki.48 Type 99 Light Bomber and Ki.49 led to dissatisfaction with the idea. A heavy weapons package was to be mounted below the bomb bay position, dramatically adding to the weight. While escort would be effectively conveyed en-route to a target, the Ki.69 would have been a drag on the return formation as the lightened bombers would be considerably faster. As a result work on the Ki.69 was terminated after the design had been fairly well finalized. The following year, on 3 February, 1943, the Hiryu airframe was officially selected for further modification as a passenger transport type with Army Air Headquarters assignment of the Ki.97 Army Experimental Transport designation. Specified to carry a crew of 2 plus 21 passengers, the Ki.97 was intended as a Ki.57-II replacement, having the same design relationship to the Ki.67 Hiryu bomber as the Ki.57-II Type 100 Transport did to the Ki.21-II Type 97 bomber. The wings, power plants, landing gear and tail surfaces of the Ki.67-Ia were retained in the design, all mounted on a new fuselage of elliptical cross section. Under construction as Mitsubishi Design Project 091107, a mock-up was completed before the JAAF terminated the assignment in 1945 in order to concentrate on defense weapons. In the original planning for the Ki.97 project, a civil transport version for use by Dai Nippon Koku K.K. (Greater Japan Air Lines) was under consideration

The first serious attempt at the modification of the Hiryu airframe as a defense fighter in its own right came late in 1943 at a time when Army Air Headquarters reviewed every possible means of producing aircraft capable of intercepting the anticipated Boeing B-29 Superfortress over Japan. Japanese intelligence had provided the JAAF with a remarkably accurate picture of the B-29's capability and configuration by the fall of 1943, and engineers at the 1st Army Air Technical Arsenal faced the need of developing equipment capable of meeting the new American bomber in combat in the spring of 1944. A rash of new interceptor fighter design projects were rushed into work, with deliveries anticipated in late 1945 or early 1946. In the interim, defense would have to be handled by modifications of existing aircraft, and the Ki.67 Hiryu was selected for further development by the JAAF engineers in November 1943 to create a night fighter defense system. The high performance, load carrying capability, and surprising maneuverability of the Ki.67 experimental aircraft, then under evaluation, led to the specification of a "Killer-Hunter" team under the respective Ki.109a and Ki.109b Army Experimental Night Fighter designations. The Ki.109a would mount two oblique upward-firing 37 mm Ho.203 cannon in the dorsal turret position while the companion Ki.109b would seek out the

Former 107th Air regiment Ki.109 in U. S. markings. The original intention was to evaluate this aircraft in tests under American supervision. The testing was canceled and the aircraft was quickly slated for destruction. USAF.

raiding bombers with a nose-mounted 40 cm. searchlight. When it became obvious that the B-29 would primarily be used as a day bomber the "Killer-Hunter" project was dropped. The Ki.109 designation was retained to be applied to a day interceptor version of the Hiryu as the Ki.109 Army Experimental Interceptor, with the assignment officially tendered to Mitsubishi in January 1944.

Major Hideo Sakamoto, in charge of the Hiryu evaluation program at the Air Examination Division at Fussa, and the man most familiar with the capabilities of the Hiryu, proposed that the Ki.67 airframe be modified to carry a standard Japanese Army 75 mm Type 88 anti-aircraft cannon in the nose. Anticipating that the long-ranged B-29s would appear over Japan without fighter escort, and that JAAF defense fighters would have the American bombers all to themselves, the radical idea was approved for development on 20 February, 1944. The muzzle velocity of the Type 88 cannon, initially developed to fire straight up at attacking bombers from the ground, would make it possible for the Ki.109 interceptor to stay out of range of the defensive fire of a B-29 formation while lobbing its heavy shells at the bombers. With that, a design team led by Mitsubishi Chief Project Engineer Kyunosuke Ozawa created a new nose and strengthened fuselage for the Ki.109, leaving the rest of the aircraft alone to retain the defensive gun positions of the Hiryu bomber.

The prototype serial 10901 was completed in August, just two months after the B-29s first appeared over Japan. With the American bombing program starting later than anticipated, and with Ki.109 development successfully moving along on its rapid schedule, it looked as if the JAAF had solved a major portion of its homeland defense problems. Tested at Kagamigahara, and quickly moved to Fussa for service evaluation, prototype serial 10901 was soon joined by a second experimental aircraft serial 10902 in October. If the Ki.109 had any problem, it was weight. As an interceptor it needed a considerable amount of advance warning in order to climb to the anticipated interception altitude. The problem was first approached by the attachment of a multi-rocket JATO package to the rear bomb bay area of the first test model serial 10901 in order to assist take-off and climb, but the idea was dropped when the aircraft became all but uncontrollable. Further weight-saving re-design took place with the third example, and first production model of the Ki.109 Interceptor, coming off Mitsubishi's production line in January 1945, having all gun positions removed except the nose cannon and a single defending 12.7 mm Type 1 Ho. 103 MG in the tail turret.

In its production form the Ki.109 had a crew of 4 consisting of pilot, co-pilot, radio operator and tail gunner, with the co-pilot acting as the nose-cannon gunner, hand-feeding the 75 mm shells to the flying field piece. 15 cannon shells were carried. JAAF orders called for 44 of the large fighters powered by Ha.104ru radial engines of 1,900 h.p., the latter a gas-turbine supercharged version of the Ha.104 used on the Hiryu bomber in order to increase the Ki.109 operating altitude. Two Ki.67-Ia aircraft, the 21st and 22nd Hiryu bombers to be produced, were used to flight test pre-production examples of the turbosupercharged engines between February and October 1944. Development of the Ha.104ru did not move along as rapidly as had been hoped, and the decision was reached to produce the first 22 examples of the Ki.l09 with the standard Ha.104 in the interim in order to get the fighter into service, with subsequent production of the remaining 22 fighters to utilize the Ha.104ru as the Ki.109-II. Experimental aircraft serial 10902 was ultimately fitted with the newer Ha.104ru radials for service evaluation while production proceeded on the first batch of Ki. 109-I fighters.

Even before the Ki.109 was being produced in its service form at Mitsubishi, formation and training of the 107th Heavy Fighter Regiment, slated to be the showcase Ki.109 unit, had begun. Established in November 1944, the 107th Regiment began its training in standard Ki.67-Ia Hiryu bombers at Micahara Air Base at Kumagaya, in December. By March 1945, when the 22nd and final Ki.109 fighter was completed, the 107th Regiment was completely organized and ready to complete its familiarization with its new equipment. In anticipation of the success of the project, a second unit was preparing to receive its allocation of the Ki.109 fighters, with deliveries to the 14th Heavy Bomber Regiment starting in May. Assigned to the Eastern Defense Sector for the air defense of Tokyo and Yokohama in the summer of 1945, the 107th Regiment was alerted a number of times but never suc-

The inglorious end of Japan's air forces. Mass destruction between October and December 1945 wiped out all aircraft remaining in Japan. Here a Ki.109 and other assorted aircraft are under the torch. Robert C. Mikesh.

cessfully intercepted a B-29 formation. With the American bombers coming in at high altitude during the daylight hours, or at low altitude at night, the Ki.109 was impotent. With the initial Ki.109-I model totally useless, and Ki.109-II development faltering, never to reach production, the project was abandoned and the 107th Heavy Fighter Regiment was disbanded on 30 July, 1945, in order to assign its pilots and crews to other more productive units. A number of Ki.109 aircraft were delivered to the Air Examination Division for evaluation as anti-invasion weapons, but the days of the Hiryu as a fighter were over. The solid-nosed "Peggy" aircraft were discovered at Fussa by occupying American forces where they were briefly considered for testing, then ultimately wrecked and burned later in the year.

The final officially designated variant of the Hiryu was authorized on 3 October, 1944, as the Ki.112 Army Experimental Convoy Fighter, reportedly a wooden adaptation of the basic Ki.67-I airframe with heavy armament. Its role may have been as a Special Attack Corps escort to protect the modified Ki.67-I two-seat Sakura-Dan Special Attack Planes from intercepting Allied fighters. The Ki.112 armament has been quoted at 8 x 12.7 mm Type 1 MGs plus a single 20 mm cannon. The Ki.112 was under development as a study project, and then dropped in the summer of 1945.

It is probable that some of the developments of the Hiryu described and identified in available records as Ki.67 variants, had their own Kitai numbers going beyond the Ki.112 designation. But the confusion of the late months of the Pacific War, the destruction of the bulk of the JAAF records only days and hours before the Allied occupation, and the lack of attention to organization in the face of JAAF project proliferation, suggests that the unknown Ki. numbers have been lost forever, if they were in fact assigned.

Had the Hiryu been available in some numbers on the perimeter of the Japanese wartime empire at the end of hostilities it would have been a powerful weapon in the hands of the many insurgencies that sprang up in the areas suddenly released from Japanese control. But it was not to be. The combined forces of low production, superior performance as a Special Attack aircraft, concentration of Ki.67 units in the home islands and the wholesale destruction of surviving examples, wiped the aircraft off of the availability list. The finest JAAF bomber of the war ended its days in flames, with aircraft piled high and torched, or briefly on static display. An operational history that might have endured for another decade came to a sudden halt. The Hiryu had a little over a year of usage and fame, and then was gone.

AIR REGIMENTS

Units operating Ki.67 Type 4 Heavy Bomber Hiryu

Regiment	When Used	Area of Operations	Former A/C	Later A/C	Comments
7th Heavy Bomber	July 1944–end of war	Philippines, Saipan, Okinawa, Manchukuo, Chosen (Korea), Japan	Ki.21, Ki.49	None	First unit to fly Hiryu in combat. Took part in Battle of Philippine Sea. Assigned to JNAF 2nd Air Fleet 25 July, 1944, as part of "T-Force" to defend Philippines. Received special training as a torpedo unit with modified Ki.67. Posted to Kanoya Naval Air Station in Oct. 1944. Removed from JNAF control June 1945. Unit moved to Kungchulung, Manchukuo and Kunsan, Chosen (Korea), later to Japan. Headquarters at Itami with units at Miho and Kameyama, Japan. Disbanded at Itami in Sept. 1945.
14th Heavy Bomber	May 1945–end of war	Japan	Ki.21	None	Unit decimated flying Ki.21 in Philippines. Reformed in Japan at Ohta and Shinden, Gumma Prefecture, May 1945 to receive Ki.67 for raids against Saipan when anticipated deliveries of newer Ki.74 "Patsy" didn't materialize. Unit disbanded at Nitta, Japan, in August 1945.
16th Light Bomber	April 1945–end of war	Chosen (Korea), Okinawa	Ki.30, Ki.48	None	Unit flew Ki.48 "Lily" light bomber in Philippines. Decimated there and reformed at Pyongyang in Chosen (Korea) to fly Ki.67. Took part in Okinawa Campaign. Unit disbanded at Heijo, August 1945.
60th Heavy Bomber	Nov. 1944–end of war	Japan, China	Ki.21	None	One of the oldest JAAF bomber units. Flew Ki.21 for six years, converted to Ki.67 for Okinawa Campaign. Headquarters at Kumamoto, Japan, with some units at Shusuishi.
61st Heavy Bomber	Nov. 1944–end of war	Singapore, Formosa	Ki.21, Ki.49	None	Decimated Ki.21 and Ki.49 unit in New Guinea. Reformed in Japan to fly Ki.67, receiving its full implement by May 1945. Unit disbanded at Kagi, Formosa, in February 1946.
62nd Heavy Bomber	Dec. 1944–end of war	Japan	Ki.21, Ki.49	None	Ki.49 unit decimated in Philippines and later reformed in Japan in December 1944 to receive Ki.67. Participated in long range over water flights to bomb B-29 bases in the Marianas. Refueled at Iwo Jima. Unit disbanded at Nishi (West) Tsukuba, north of Tokyo, at end of war.

Unit	Dates	Location	Other Aircraft	History
74th Heavy Bomber	May 1945-end of war	Formosa, Japan	Ki.49	None	Decimated Ki.49 unit in the Philippines, active in suicide missions. Reformed in Japan with Ki.67 as part of Ten-Go suicide operation, retaining some Ki.49 for backup and training. Incorporated into 74th Heavy Bomber Regiment after the fall of Okinawa. Disbanded at Matsumoto, Japan, at end of war.
98th Torpedo	May 1944-end of war	Japan, Okinawa, Chosen (Korea)	Type I (Fiat), Ki.21	None	First unit to fly Ki.67 in Bomber company with 7th Heavy Bomber Regiment under JNAF control as part of "T-Force" in Battle of Philippine Sea. Ki.67 modified to torpedo bombers with special training at Fussa. Assigned to JNAF 2nd Air Fleet. Removed from JNAF control July 1945. Unit disbanded at end of war at bases at Kodama, Tachiarai and Shinkawa.
107th Heavy	10 Nov. 1944-30 July 1945	Japan	None	Ki.109	Heavy bomber Bomber unit formed at Hamamatsu in November 1944. Mission converted to home island defense unit training in modified Ki.67 for conversion to Ki.109 Heavy Fighter. Based at Micahara, Kumagaya, Japan. Disbanded at Hamamatsu 30 July 1945.
110th Heavy Bomber	16 Oct. 1944-end of war	Hamamatsu, Japan, Iwo Jima, Okinawa, Chosen (Korea)	None	None	Home island defense unit flying Ki.67. Formed in Japan at Hamamatsu with Ki.67 as original equipment. Established for long range over water flights to attack American airfields in the Marianas. Became part of Training Air Force 7 Nov. 1944, attacking Saipan 6 Dec. Flew from Hamamatsu to refuel at Iwo Jima and on to B-29 bases. Took part in Okinawa Campaign. Reassignment to Kengun, Kyushu, and Kinshu, Chosen, as operational unit underway at end of war. Unit disbanded at Kumamoto in August 1945.
170th Heavy Bomber	7 Aug. 1945-end of war	Japan	None	None	Last unit to receive Ki.67. 170th Air Regiment formed from remnants of 60th and 100th Air Regiments formed on 7 August 1945, eight days before end of war. Single aircraft sortie on 8 August 1945.

Company	When Used	Area of Operations	Former A/C	Later A/C	Comments
28th	25 Feb. 1945-end of war	Malaysia	Ki.54	None	Heavy bomber training unit to convert to regiment. Unit formed at Ipoh, Malaysia, with Ki.67 as original equipment with Ki.54 for multi-engine and flight engineer training.

TRAINING SCHOOLS

School	When Used	Area of Operations	Former A/C	Later A/C	Comments
Hamamatsu Heavy Bomber Instructing Flight Division	20 June 1944-10 July 1945	Hamamatsu, Japan	Ki.21	Ki.74	Leading Army Heavy Bomber school. Only Flight school to provide Ki.67 pilot and crew training for unit formation.
Tokorozawa Instructing Maintenance Division	June 1944-end of war	Tokorozawa, Japan	All current JAAF aircraft	All current JAAF aircraft	Maintenance and repair training for units flying Ki.67 Hiryu, Ki.109 Heavy Fighter and other variants.

MISCELLANEOUS ASSIGNMENTS

Unit	When Used	Area of Operations	Former A/C	Later A/C	Comments
Army Air Test Department	Jan. 1943-end of war	Fussa, Japan	All current JAAF aircraft	All current JAAF aircraft	Tested Ki.67 experimental models with extensive development and evaluation of torpedo and Special Attack versions. Testing under direction of Major Hideo Sakamoto.
2nd Independent Flying Unit	July 1944-Dec. 1944	Hamamatsu, Japan	None	None	Special mission Ki.67 unit established at Hamamatsu Heavy Bomber Instructing Flight Division to attack American B-29 bases on Saipan. Aircraft and mission were incorporated in 110th Heavy Bomber Regiment at end of 1944.

SPECIAL ATTACK UNITS

Unit	When Used	Area of Operations	Former A/C	Later A/C	Comments
Fugaku Special Attack Corps	25 Oct. 1944-Jan. 1945	Philippines, Formosa	None	None	Fugaku (Mt. Fuji) Special Attack Corps Company formed at Hamamatsu with 26 men and 12 highly modified Hiryu aircraft assigned to Special Attack during Philippines campaign. First use 7 Nov. 1944. Last on 12 Jan. 1945. Later modified Ki.67 aircraft equipped with Sakura-Dan (German "Cherry Bomb").

Mitsubishi Ki.67/Ki.109 HIRYU

Unit	When Used	Area of Operations	Former A/C	Later A/C	Comments
74th Heavy Bomber	May 1945–end of war	Philippines, Japan	Ki.49	None	Special Attack units of the 74th regiment officially formed 14 Dec. 1944. Flew Ki.49 on suicide raids in Philippines, to be decimated there. Reformed in Japan with Ki.67 as part of Ten-Go homeland defense suicide operation May 1945.
95th Heavy Bomber	May 1945–end of war	Philippines, Japan	Ki.49	None	Special Attack units of the 95th regiment officially formed 14 Dec. 1944. Flew Ki.49 on suicide raids in Philippines, to be decimated there. Reformed in Japan with Ki.67 as part of Ten-Go homeland defense suicide operation May 1945.
Unnamed Tokko Unit	May 1945	Japan	None	None	Suicide attack on Okinawa 24 May 1945. Formed by 60th Air Regiment.
Unnamed Tokko Unit	April 1945–May 1945	Japan	None	None	Formed by 62nd Air Unit Regiment flying standard Ki.67 and Ki.67 Sakura-Dan Special Attack planes. Attacks against shipping east and west of Okinawa 7 April 1945 and 25 May 1945.
Unnamed Tokko Unit	April 1945	Chosen (Korea)	None	None	Formed by 110th Air Unit Regiment. Attack in Okinawa area 26 April 1945.
Shichisho Kojun Unit ("Everlasting Shield of the Emperor")	Jan. 1945	Malaya	None	None	Formed by 1st Field Air 2nd Flying Replacement Unit at Syonan (Singapore). Attack on shipping southwest of Sumatra 29 Jan. 1945.
Shichisho Shinrai Unit ("Everlasting Divine Thunder")	June 1945	Surabaya, Java	None	None	Formed by 61st Air Regiment at Surabaya Airfield, Java. Attack at Balikpapan, Borneo, 25 June 1945.

NAVAL AIR GROUPS

Air Group	When Used	Area of Operations	Former A/C	Later A/C	Comments
762nd Code: 762 2nd Air Fleet	Aug. 1944–Jan. 1945	Kyushu, Japan, Philippines	G3M, G4M	P1Y	First naval unit to fly JAAF bomber type in combat. Took part in Battle of Philippine Sea in engagements south of Formosa 12/14 Oct. 1944 in company with JAAF 7th and 98th Regiments flying torpedo aircraft. All units were under command of JNAF 2nd Air Fleet as Yasukuni Naval Air Group.

Yokosuka Naval Air Test Center Code: Ko	April 1943-July 1944	Yokosuka, Honshu	G3M, G4M, G5N, P1Y G8N	Testing of modified Ki.67 evaluation aircraft serials 6717 and 6718 with torpedo racks. Base located at Yokosuka at Kanagawa on Tokyo Bay, Honshu, Japan.

Note: These lists are not to be regarded as complete as only those units for which JAAF Ki.67 and JNAF (Ki.67) Yasukuni use has been confirmed have been identified.

Units operating Ki.109 Interceptor Fighter

AIR REGIMENTS

Regiment	When Used	Area of Operations	Former A/C	Later A/C	Comments
14th Heavy Bomber	May 1945-June 1945	Japan (Home Island Defense)	Ki.67	None	Backup Ki.109 Interceptor Fighter unit in event the 75mm cannon Heavy Fighter was successful in B-29 interceptions. Lack of success led to abandonment of defense fighter mission.
107th Heavy Fighter	10 Nov. 1944-30 July 1945	Japan (Eastern Defense Sector)	Ki.67	None	Assigned to B-29 interception for Tokyo and Yokohama air defense. No successful interceptions accomplished. Unit disbanded at Micahara, Kumagaya, Japan, in July 1945 after lack of unit success. Pilots and crews assigned to other regiments.

TRAINING SCHOOLS

School	When Used	Area of Operations	Former A/C	Later A/C	Comments
51st Flight Training Division	10 Nov. 1944-8 Feb. 1945	Japan (Eastern Defense Sector)	Ki.67	None	Heavy Fighter operations of 107th Fighter Regiment under command of 51st Flight Training Division to train and expand potential of Ki.109 Experimental Interceptor Fighter. Venture not successful.

MISCELLANEOUS ASSIGNMENTS

Unit	When Used	Area of Operations	Former A/C	Later A/C	Comments
Army Air Test	Oct. 1944-end of war	Fussa, Japan	All current JAAF aircraft	All current JAAF aircraft	Numerous Ki.109 Department experimental models tested at Fussa for Interceptor Fighter version as well as development of Ki.109-II version.

Note: These lists are not to be regarded as complete as only those units for which JAAF Ki.109 use has been confirmed have been identified.

SPECIFICATIONS: Mitsubishi Ki.67 Type 4 Heavy Bomber Hiryu (Flying Dragon)

Note: All dimensions in original Japanese metric. Dimensions and climb in meters (m.), weights in kilograms (kg.), distance in kilometers (km.) and speeds in kilometers per hour (km./hr.). Data in parenthesis are estimates or approximate.

Model and Specs	Ki.67 Exper.*	Ki.67 Eval.**	Ki.67-Ia	Ki.67-Ia Torpedo****	Ki.67-Ib	Ki.67-Ia L/R	Ki.67-Ic	Ki.67-I S/A	Ki.6 S/A
Span (m.)	22.50	22.50	22.50	22.50	22.50	24.00	22.50	22.50	22.5
Length (m.)	18.70	18.70	18.70	18.70	18.70	18.70	18.70	18.70	18.7
Height (m.)	(8.30)	7.70	7.70	7.70	7.70	7.70	7.70	7.70	7.70
Wing Area (m.2)	65.85	65.85	65.85	65.85	65.85	65.85	65.85	65.85	65.8
Weight Empty (kg.)	--	--	8,649	8,649	--	--	--	--	--
Weight Loaded (kg.)	--	--	13,765	13,765	--	--	--	--	--
Weight Loaded Max. (kg.)	--	--	--	--	--	--	--	--	--
Max. Speed (km./hr.)	537/6,090	--	537/6,090	537/6,000	--	--	--	--	--
Cruising Speed (km./hr.)	400/8,000	--	400/8,000	400/8,000	--	--	--	--	--
Climb (m./min.)	--	--	6,000/14'30"	6,000/14'30"	--	--	--	--	--
Armament-M.G. (mm.)	3x7.92, 2x12.7	2x7.92, 2x12.7	4x12.7	4x12.7	5x12.7	4x12.7	5x12.7	--	--
Armament-Cannon (mm.)	--	1x20	2x20	1x20	1x20	1x20	1x20	--	--
Armament-Bombs (kg.)	500	500	500/800	1070 Torpedo	750	500	1250	2x800	1x2
Engines	2	2	2	2	2	2	2	2	2
Power-Mfr.	Mitsubishi	Mitsubishi	Mitsubishi	Mitsubishi	Mitsubishi	Mitsubishi	Mitsubishi	Mitsubishi	Mits
Type	Ha.104	Ha.104	Ha.104	Ha.104	Ha.104	Ha.104	Ha.104	Ha.104	Ha.
H.P.	1,900	1,900	1,900	1,900	1,900	1,900	1,900	1,900	1,90
Crew	6	8	8	8	8	6/8	6/8	3	2
Aircraft-Mfr.	Mitsubishi	Mitsubishi	Mitsubishi	Mitsubishi	Mitsubishi, Kawasaki, Nippon Kokusai, Rikugun	Mitsubishi	Mitsubishi	Mitsubishi	Mits
First Built	Nov. 1942	May 1943	Nov. 1943	Sept. 1944	Dec. 1944	Spring 1945	(Sept. 1945)	Sept. 1944	Feb. 1945)
Number Built	3	16	431***	271†	267††	12†	None	15†	2†
Serials	6701-6703	6704-6719	--	--	--	--	--	--	--

*	Projected engines were Ha.101, Ha.103 and Ha.104. Ha.104 selected.
**	Evaluation aircraft 6717 and 6718 modified to carry torpedo; 6716 and 6717 fitted with Ha.214.
***	Two examples powered by Ha.104ru for Ki.109 testing. Others became glider tugs, I-GO-A Guided Missile launch aircraft, searchlight, radar, etc.
****	Modified to carry torpedo. Also used by JNAF as Yasukuni.
†	Included in Ki.67-Ia production.
††	Mitsubishi 156; Kawasaki 81; Nippon Kokusai 29 from Mitsubishi parts; Rikugun 1; total 267.
†††	German "Cherry Bomb" Sakura-Dan.
††††	Military and civil versions projected. Would have been postwar civil transport.
§	Not completed.
§§	Prototype serial 10901 experimentally fitted with JATO rockets.
§§§	Second airframe serial 10902 fitted with supercharged Ha.104ru.
§§§§	Projected production. Work began 3 October 1944. None completed.

Ki.67-II	Ki.69 Escort	Ki.97 Transport	Ki.109a "Killer"	Ki.109b "Hunter"	Ki.109-I Exper.	Ki.109-I	Ki.109-II Prototype	Ki.109-II	Ki.112 Fighter
22.50	22.50	22.50	22.50	22.50	22.50	22.50	22.50	22.50	22.50
18.70	18.70	20.00	--	--	17.95	17.95	17.95	17.95	18.70
7.70	7.70	5.70	7.70	7.70	7.70	5.80	5.80	5.80	7.70
65.85	65.85	67.50	65.85	65.85	65.85	65.85	65.85	65.85	65.85
--	--	8,450	--	--	--	7,424	--	--	--
--	--	13,000	--	--	--	10,800	--	--	--
--	--	--	--	--	--	--	--	--	--
--	--	546/5,400	--	--	--	550/6,090	--	--	--
--	--	400	--	--	--	--	--	--	--
--	--	--	--	--	--	--	--	--	--
5x12.7	--	2x12.7	--	--	-	1x12.7	1x12.7	1x12.7	(8x12.7)
2x20	--	--	2x37	Searchlight	1x75	1x75	1x75	1x75	(1x20)
750	--	--	--	--	--	--	--	--	--
2 Mitsubishi Ha.214 2,400	2 Mitsubishi Ha.104 1,900	2 Mitsubishi Ha.104 1,900	2 Mitsubishi Ha.104 1,900	2 Mitsubishi Ha.104 1,900	2 Mitsubishi Ha.104 1,900	2 Mitsubishi Ha.104 1,900	2 Mitsubishi Ha.104ru 1,900	2 Mitsubishi Ha.104ru 1,900	2 Mitsubishi Ha.104 1,900
7/9 Mitsubishi	6 Mitsubishi	2+21 pass. Mitsubishi	4 Mitsubishi	4 Mitsubishi	4 Mitsubishi	4 Mitsubishi	4 Mitsubishi	4 Mitsubishi	(6) Mitsubishi
(Sept. 1945)	Dropped	Dropped	Dropped	Dropped	Aug. 1944	Jan. 1945	Feb. 1945	(Sept.	Dropped
2§	None	None††††	None	None	3§§	22	1§§§	(22)§§§§	None
--	--	--	--	--	--	--	--	--	--

Note:
- Exper. - Experimental
- Eval. - Evaluation
- S/A - Special Attack
- L/R - Long Range

**Mitsubishi/Nakajima G3M1/2/3 96 RIKKO L3Y1/2
in Japanese Naval Air Service**

Richard M. Bueschel

Richard Bueschel revises and updates his classic series of books on Japanese Naval and Army Air Force aircraft of World War II. The Japanese navy Mitsubishi/Nakajima G3M1/2/3 96 RIKKO (Nell) is presented in this volume. All variations and markings are covered in this sixth book in a multi-volume series.

Size: 8 1/2" x 11", over 100 b/w photographs
64 pages, soft cover
ISBN: 0-7643-0148-9 $14.95